A Treasure Hunting Text

Bob Moore

360 – 770 – 1171

Ram Publications
Hal Dawson, Editor

Modern Treasure Hunting

> *The* practical guidebook to today's metal detectors; "how-to" manual that explains the "why" of modern detector performance.

Treasure Recovery from Sand and Sea

> Tells any treasure hunter how to reach the "blanket of wealth" beneath sands nearby and under the world's waters.

Modern Electronic Prospecting

> Explains in layman's language how anyone can use a modern detector to find gold nuggets and veins of precious metal.

Buried Treasure of the United States

> Complete field guide for finding treasure; includes state-by-state listing of sites where treasure can be found.

Modern Metal Detectors

> Advanced handbook for field, home and classroom study to increase expertise and understanding of metal detectors.

Successful Coin Hunting

> The world's most authoritative guide to finding valuable coins with all types of metal detectors.

Getting the Most from your Grand Master Hunter

> Specific instructions for getting "plus" performance from Garrett's new computerized microprocessor-controlled detector.

Treasure Hunter's Manual #6

> Quickly guides the inexperienced beginner through the mysteries of full time treasure hunting.

Treasure Hunter's Manual #7

> The classic book on professional methods of research, recovery and disposition of treasure.

Weekend Prospecting

> Describes how to enjoy holidays and vacations profitably by prospecting with metal detectors and gold pans.

Gold Panning is Easy

> Excellent field guide shows the beginner how to find and pan gold as easily as a professional.

Treasure Hunting Pays Off

> An introduction to all facets of treasure hunting . . . the equipment, the targets and the terminology.

The Secret of John Murrell's Vault

> Fictional story of a search that really took place for a treasure still awaiting discovery. A True Treasure Tale.

Garrett Guide Series — Pocket-size field guides

> *Find an Ounce of Gold a Day*
> *Metal Detectors Can Help You Find Wealth*
> *Find Wealth on the Beach*
> *Metal Detectors Can Help You Find Coins*
> *Find Wealth in the Surf*
> *Find More Wealth with the Right Metal Detector*

Modern Electronic Prospecting

by Roy Lagal
and Charles Garrett

RAM
BOOKS

ISBN 0-915920-58-1
Library of Congress Catalog Card No. 88-061452
Modern Electronic Prospecting
©Copyright 1988
Charles L. Garrett & Roy Lagal

First Edition Printing. September 1988

88 89 90 9 8 7 6 5 5 4 3 2 1

For FREE listing of RAM treasure hunting books write
Ram Publishing Co. • P.O. Box 38649 • Dallas, TX 75238

Book and cover designs by Mel Climer
Cover photographs by Hal Dawson

Contents

Editor's Note

For almost a decade *the* guidebook for seeking gold and other precious metals with a metal detector has been *Electronic Prospecting* by Lagal and Garrett. Prospectors in the field recognized it as a book written exclusively for them. The authors knew their subject well and were able to describe it in a manner adequate for both the long-time prospector and the beginner.

From the knowledge gained by a lifetime of experience Roy Lagal knows how a prospector thinks and what he is looking for. This expertise was evident in the first book. As one of the world's leading developers of the modern metal detector – and an active prospector in his own right – Charles Garrett understands how a metal detector can help both amateurs and professionals find gold. His expertise was likewise vital to this book's success.

And, it is true that *Electronic Prospecting* was an instant success and remained successful. It underwent numerous reprintings and its text was continually updated. Roy and Charles demanded that changes be made because new equipment literally obsoleted the original text time and again. They refused to offer incomplete or misleading information to users of metal detectors.

Much new literature has been added since *Electronic Prospecting* was first printed. Still, their book has retained its claim as the leader in the field.

Recent developments, particularly in computerized metal detectors, caused Roy and Charles to review carefully the current version of their best seller. They agreed that a totally new book was needed – not just a revision.

This new book is the result. Those of you who loved the original book and have come to memorize passages from it will find much here that is familiar. The laws of physics and the facts of history have not changed. Yet, you will also find that most of this book is new, particularly the material regarding the use and application of newly developed gold hunting equipment. Here, then, is *Modern Electronic Prospecting* . . . a book written for the prospectors of the 1990's and beyond who plan to use the latest electronic metal detecting equipment available.

Autumn 1988   *Hal Dawson, Editor*
                                                 Ram Books

About the Authors

Charles Garrett and Roy Lagal possess more knowledge between the two of them about finding gold with a metal detector than any other pair of individuals in the world today. *This statement is indisputable.* Garrett and Lagal scarcely need an introduction to any electronic prospector.

Roy Lagal is a lifetime prospector who has "put his money where his mouth is" time and time again. His livelihood has depended upon his prowess with a metal detector and gold pan. He has achieved success as a full-time prospector in the United States, Mexico and Canada for some 40 years. For the past 20 years he has worked with Charles Garrett and others to transform the field of electronic prospecting into a science. He is the designer of the "Gravity Trap" gold pan, and with books such as *Weekend Prospecting, Gold Panning Is Easy,* and *Find An Ounce of Gold A Day* he has elevated the hobby of prospecting to a professional level that almost anyone can achieve.

Charles Garrett has been a treasure hunter since the 1940's and founded one of the most successful metal detector companies in the industry more than 25 years ago. A graduate engineer, he introduced discipline to the manufacture of metal detectors generally and to the field of electronic prospecting specifically. He has searched for gold on every continent except Antarctica and has written of his experiences in numerous books, some in collaboration with Roy Lagal. Several of his books such as *Modern Metal Detectors, Successful Coin Hunting* and *Treasure Recovery from Sand and Sea* have become accepted as textbooks for the hobby of metal detecting.

Modern Electronic Prospecting represents an effort by these two experts to share with both professional and amateur prospectors the knowledge they have developed concerning electronic prospecting and the most recent advancements in metal detection equipment.

The Editor

Chapter 1

Gold's Lure and Lore

G old has been precious to man since the dawn of time. As an ornament, as a symbol of wealth, as a means of barter, the yellow metal has played a vital role in world history. Kingdoms were won and lost because of gold. New lands were sought and conquered. The value of gold and the quest for it helped draw the map of the entire Western Hemisphere and explains why Spanish is spoken almost exclusively between the Rio Grande River and the Antarctic Ocean while the natural resources of most of North America accrued to English-speaking peoples.

The story of the Ancient World was indeed written in gold. Even when it was too scarce for daily transactions, the precious metal was a symbol of wealth for the ancient civilizations – in China from about 1200 B.C.; Egypt, 1000 B.C.; and in Babylon and Minoa from the 3d century B.C. Then in the 16th century gold led the greatest conquest of new lands that will ever be seen in the history of our planet as the Spanish plundered ancient civilizations in Mexico and Peru to enrich the royal storehouse in Madrid.

Three centuries later the search for gold and other precious metals hastened the development of the young United States of America and played a significant role in both its boundary lines and its destiny. The search for gold continues to this day.

Countless tons of gold have been taken from the earth by professional miners, using expensive equipment and sophisticated methods. Cursory study of geography, geology and chemistry reveals, however, that most of the earth's gold is still waiting to be recovered. In 1849, the year that the discovery at Sutter's Mill began the great California Gold Rush, gold production was 11,866 ounces. Production climbed steadily, reaching a

1

peak of 2,782,018 ounces in 1856. Nine years later production had declined below a million ounces. The great California Gold Rush was over. Miners and prospectors had moved on to the greener – and more golden – pastures of Colorado, Alaska and Australia.

Even with surface gold removed from the Mother Lode country of California, numerous geological surveys and studies have suggested that only some 15 to 20% of the gold in California has actually been recovered. Based on this data, it seems obvious that a vast amount of gold still remains to be discovered . . . not only in California, but in all other parts of the world.

Gold is not present in the area where you live, you say? Perhaps no one has looked hard enough for it!

20th Century Gold Rush

For more than four decades from 1933 until the end of 1974 citizens of the United States were prohibited by law from owning gold. This ban was among the measures inaugurated by President Franklin D. Roosevelt in the early days of his presidency as he sought to shore up the economy of the nation at the depths of the Great Depression. Throughout this period when private ownership was prohibited, the government set the price of gold at artifically low prices, $35 per ounce in 1974.

After the ban was lifted, the price of gold soared. In the harrowing times of steep inflation and high interest that followed, this price exceeded $800 per ounce. Such a sharp increase in value, coupled with man's adventurous spirit, caused a small scale gold rush back to the 19th century camps and ghost towns of the American West. Metal detector manufacturers benefited commensurately. With the Garrett factory in full production, deliveries lagged some three months behind sales during the boom.

Although the price of gold has decreased significantly from its highs of the early 1980's, it remains – as this book is written – some 12 times greater than it was when the ban on ownership was lifted in 1974. Furthermore, there is no doubt that a strong demand for gold exists in the private sector. With industrial use of gold steadily increasing, prices can be expected to rise generally – especially in relation to international economic conditions.

Just as the 19th century prospectors sought to control their own fate and fortune through the discovery of nature's wealth, so do today's seekers after gold. The spirit of the 49'ers and of the Klondike prevails among both professional and recreational prospectors who have discovered the pleasures and profits of searching for gold.

Who Is Prospecting?

Doctors, attorneys, businessmen, students, senior citizens . . . entire families . . . from all walks of life and all income levels have found that the healthful, relaxed outdoor life of weekend or vacation prospecting can yield big dividends—in dollars and cents as well as pure pleasure. A single ounce of gold, in many cases, is worth the equivalent of several days' pay. One fair-sized nugget can be worth more than a month's salary. These "instant riches" are obviously part of the attraction of the business/hobby known as recreational mining.

Men and women, boys and girls who are searching for gold work at regular jobs, but on weekends, holidays and vacations they join families and friends to head for one of the many thousands of areas open to the public where gold can be found. They set up camp by a stream, then use a pan or dredge along with their metal detector to find gold. Or, they may travel to an old ghost town or deserted mining camp and search for nuggets or ore veins or valuable mineral/ore specimens overlooked by early day prospectors who operated without the benefit of electronic equipment.

Modern Equipment

Three developments have greatly increased the ability of the recreational miner to hit paydirt in comparison with the sourdough prospectors who flooded California in the 1850s and went to Alaska at the turn of the century:

1. Availability of an easy-to-use, highly efficient gold pan;
2. Production of lightweight, portable dredges that can be effectively operated by one or two people;
3. Development of the metal detector. This single tool makes locating precious metals simpler for *all* prospectors, no matter what their level of experience, no matter where they are searching . . . on land or in the water. The earliest metal detec-

tors were welcomed just after World War II by prospectors who sought any advantage that would improve their chances of striking it rich. Today's rugged, yet highly sensitive, ground canceling metal detectors are capable of detecting precious metals as they operate in even the difficult terrain of the most highly mineralized rocks or soil.

Chapter 2

Where to Find Gold

It can be said truthfully that gold can be found in nature throughout the United States – in all 50 states. An immediate postscript must be added: while traces the precious yellow metal may be *physically present* in all of the states, *sufficient quantities* to make prospecting profitable occur in only about half of the states. In the others gold is so fine and in such minute quantities that its presence is a fact of chemistry rather than of potential wealth. Recovery is not only impractical here but virtually impossible . . . especially for the recreational prospector.

Contrary to prevailing opinion, the major gold producing states are not concentrated in the Rocky Mountains or the states west of them. States where gold can be found and recovered effectively are located from ocean to ocean and from border to border. The major gold production states, listed roughly in their order of total production:

California	Arizona	South Carolina
Colorado	Oregon	Tennessee
South Dakota	Idaho	Virginia
Alaska	Washington	Alabama
Nevada	New Mexico	Texas
Utah	North Carolina	Michigan
Montana	Wyoming	Wisconsin
	Georgia	

Production from these states has amounted to more than 300 million ounces – or well over a trillion dollars at today's price. It is obvious that gold production in the United States has been no "nickle-and-dime" matter, and production continues at this moment in many of the above states.

First, look at the list; next, locate on the accompanying map the areas of production that are closest to you. Then, after a little research, using the Appendix of this book, your public library and other sources that you might discover . . . off you go into the gold fields. But, first, your research!

Do Some Investigating!

Once you have decided upon the area(s) in which you wish to prospect, check with the governmental agencies and request information and/or literature that would be of assistance to you. Federally owned lands, located generally in the Western states, are under the supervision of the Bureau of Land Management. Their regional offices which are listed in this book offer valuable information and pamphlets on staking claims and other matters that are important to prospectors.

Most states have some sort of Bureau of Mines or Geological Surveys which can give you information on the incidence of gold is particular areas. They can also be helpful in providing information on prospecting geology and gold recovery. A listing of the addresses of the Bureaus of Mines in the Western states is included in the Appendix. Seek out the Bureau of Mines or Geology in your state. You might be surprised at the information they can provide on prospecting with a metal detector.

Still another source of information are the appropriate state Bureaus of Tourism. As the value of tourists has come to be appreciated more and more, many of the states have prepared maps and other materials showing the locations of gold deposits, gem fields, ghost towns, mining districts and other points that

Facing
Roy Lagal observes as Charles Garrett hunts ore veins or nuggets with the deepseeking Bloodhound Depth Multiplier that can look far below the surface.

Over
This beautiful golden nugget was found with a metal detector in the Outback of Western Australia near the town of Kalgoorlie.

might interest the electronic prospector.

Go to public and school libraries. Look for all those books and other research sources on the gold districts in which you plan to work. Find out about the prospecting that has already been carried out in this area; it may give you a better indication of your chances. Always remember, you will be more successful working old mines that produced free milling gold than those which required complicated chemical processing or leaching techniques to recover the metal. Find out what kind of gold was produced in the areas you investigate.

Are there dredge tailing piles to work? Are there good placer areas where you may find nuggets? Are you going to search for rich gold ore veins? Decide *precisely* what type of prospecting you are going to do, and then equip yourself accordingly.

If possible, talk to other prospectors who have worked the areas you intend to enter. There is absolutely no question that your chances of success will increase in proportion to your familiarity with an area. And, proper equipment will enable you to be more effective in your recovery efforts. Know *where* you plan to work before you ever take to the gold fields, and you will have a far better chance of returning with some good specimens, be they placer, nuggets or ore samples.

Further information is available from the other books on prospecting published by Ram Publishing. Current titles are included elsewhere in this book. For a compete list write Ram at P. O. Box 38649, Dallas, TX 75238.

Facing
A prospector's knowledge of terrain is important since the modern metal detector can report only what is beneath its searchcoil.

Over Top
Gold comes from the earth in many sizes and shapes, but every piece of it is beautiful. All of these nuggets were found with modern metal detectors.

Bottom
After successful gold panning a prospector uses the "Gold Guzzler" suction bottle to separate his nuggets from black sand which he will also save.

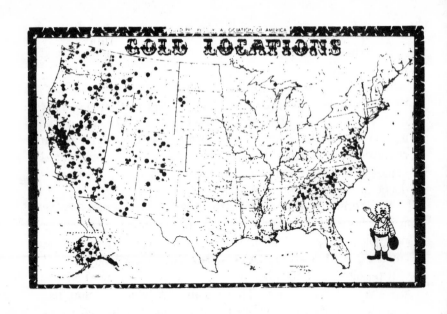

GOLD LOCATIONS

Chapter 3

Recovery Methods

G old is where you find it! And, you must *find* gold before you can begin to solve the age-old mysteries of recovering it. Some areas naturally are going to produce more "color" than others. You should always remember that gold is a *heavy* metal and will tend to follow the shortest path as it moves along in a stream or river. When you study flowing water and find a long stretch where it moves rapidly, be alert for cracks, crevices or other anomalies on the bottom. They may have stopped the movement of the gold in years and centuries gone by and continue to trap it. At the end of such long stretches remember always to search for gold on the *inside* of bends in the river. Since gold will always seek the shortest distance in its downhill path, it will "cut across" wide turns. Check also behind boulders and gravel bars where gold might have settled when the water that was carrying it along slowed down.

Many obvious locations have produced great amounts of gold in the past and continue producing to this day. But, remember that conditions change, especially over hundreds and thousands of years. Large deposits of gold may have been left by nature in places that look highly unlikely today. Gold is where you find it!

The Gold Pan

Believe us! Regardless of whether you intend to recover gold by locating it with a metal detector, capturing it in a dredge or by sluicing, a gold pan will remain your *primary tool*. Pans will obviously be necessary in streams, but you will also find them useful far away from water, often in ways that the amateur prospector could never imagine. Dry panning is sometimes the only practical way of discovering gold in desert areas. The pan may also be used for easy recovery of metallic targets signaled by your detector . . . objects that might be difficult or impossible to locate otherwise.

Simply stated, if you intend to seek gold in the field, it is vital that you not only have a good gold pan but that you understand its importance and know how to use it properly.

Just as important here is having the *right* kind of pan. Of course, any kind of basin bowl will do. Remember the old prospector from the Western movies riding his decrepit burro into a beautiful Hollywood sunset? All he had was a skillet or pie pan with which he scooped up gravel and panned for gold. As far as the pan was concerned, this depiction is still moreorless accurate. But, pan designs have improved greatly since the old prospector's time. Today's gold pan is lighter in weight and offers greater speed in testing and classifying concentrates. It is also easier to handle and provides safer, surer results, especially for the beginner. Absolutely no experience is required for a beginner to enjoy success in the first panning session.

During the gold rush days of the 19th and early 20th centuries, gold panning was more than hard work; it was back-breaking labor. Unless a panner was lucky, it was usually not especially profitable. Today, however, gold panning is much easier and far more productive. This has happened not only because of the increased price of gold but because of the modern gold pans that are now available.

Of course, we believe the finest and most effective gold pan today to be the "Gravity Trap" pan. Invented by Roy Lagal **(U.S. Patent #4,162,969)** and manufactured by Charles Garrett's company, Garrett Electronics, its effectiveness has been proven by worldwide success and acceptance. Made of unbreakable polypropylene, the pan is far lighter and easier to handle than the old metal pans. More importantly, the Gravity Trap pan has built-in gold traps in the form of sharp 90-degree riffles. These riffles are designed to trap the heavier gold and allow fast panning off of unwanted sand, rocks and gravel.

The pan is forest green in color which has been proved in extensive laboratory and field tests to show gold, garnets, precious gems and black sand better than other colors, including black. After only a little practice, a weekend or recreational placer miner using this new pan can work with equal or greater efficiency than the most proficient professional using the old style metal pans or those of black plastic design.

Today's recreational miner can achieve excellent results by using a good ground cancelling metal detector to locate gold deposits, then panning them with a Gravity Trap gold pan. Whether gold is found in profitable quantities or not, the pleasure of sitting at the edge of a mountain stream or in a long-forgotten dry gulch is one that should not be overlooked. Meanwhile, the panner is seeking to produce income with two bare hands, knowing full well that the chance always exists of hitting "the big one."

Since Gravity Trap pans can be used for both wet and dry panning, even old stream beds and washes can be made to produce gold. Built-in riffle traps can be depended upon to trap gold whether water is present or not. True, dry panning is more difficult than wet panning and requires more practice. It can be sometimes more profitable, however, because dry streams that have not seen water for many years – or centuries – can sometimes be especially productive since they were probably passed by during the busier gold rush days. Old timers, remember, with less efficient metal pans, preferred to work with running water because panning there was easier. You may be the first person to ever pan for gold in that specific location. That fact alone can make a trip to the gold fields worthwhile.

Lightweight Dredges

Crude gold dredges, ugly in appearance and often bigger than many houses, have been used for gold recovery since before the turn of the century. Dredging is by far the easiest way to retrieve gold from river and stream gravels. Yet, the tools of dredging have been improved for the weekend or recreational prospector just like the gold pan. Today's modern lightweight suction dredges are capable of working bedrock cracks and crevices almost with the speed and efficiency of an underwater vacuum cleaner. And, they are simple enough to be operated by a single person.

Powered by a small gasoline engine which drives a centrifugal pump to create the vacuum needed to suck up sand, rock, gravel and other gold-bearing materials from stream beds, dredges for today's recreational miner are extremely efficient and light enough to be packed into wilderness locations by one man. The sluice box of the dredge is designed to retain even fine gold, yet spill large rocks and other materials back into the water without clogging the riffles. Ease of cleanup depends upon the size and type of dredge used.

Dredge operation is simple. The suction hose, which can range from 1 1/2 to 6 inches in diameter, is worked slowly along the bottom of a stream. It chews away at a gravel bank, reaching ever deeper toward bedrock. Some gold can be recovered from the gravel, but the richest deposits are usually trapped in crevices of the bedrock itself. Suction is usually sufficient – with the use of a creviving tool – to break up jammed gravel in cracks and to clean the rich deposits of fine gold and nuggets out of the crevices. The material is pulled up through the hose and dumped onto the riffle box. Flow of water keeps sand, rocks and lighter material moving along the riffles until they drop out of the opposite end of the box into the water. Heavier gold flakes and nuggets are trapped behind the riffles where they remain until the concentrates are panned down, usually at the end of the day or dredging session.

Before you use a suction dredge in a river or stream, check with local authorities concerning its legality. Some states require a dredging permit. In California it is obtainable at the local office of the Department of Fish and Game. Other states prohibit dredge operation in certain locations.

This dredge ban was enacted because the huge, old dredges moved great mountains of sand and gravel. Some irresponsible operators left ugly scars that are still visible on the countryside. It is indeed unfortunate that their lack of concern for our environment should prohibit a recreational miner from using a small dredge today since the smaller models cause no ecological harm. Indeed, all material removed from a streambed is returned immediately to it! In fact, this turnover of the stream's bottom serves a worthwhile purpose since it provides new supplies of food for fish and other forms of life.

Operations of the gigantic old dredges often left huge stacks of rocks and mud piled alongside streams, producing not only an eyesore but also, in some cases, actually changing the course of the stream itself. Always remember, however, that these eyesores can lead you to wealth that can be discovered with your metal detector. Oftentimes, solid gold nuggets and gold enclosed in large mudballs passed through the dredges and were dumped out with tailings. Your ground cancelling metal detector can easily locate this gold missed by the early day dredgers.

Hunting Float, Pockets

We've already learned that "float" can be defined as ore ... away of somehow from its original bed and was carried ... (usually down hill by gravity, wind, water, earth move-ment or other acts of nature). Float is the found lodged in hollow places and behind boulders and other barriers to its downhill movement. These heavy pieces of ore form pocket concentrations and have often been covered by an overburden of rock and gravel that has washed in on top of these deposits.

Modern ghost town mineral deposits ... you locate these rich pockets of metallic ores. Medium to large ore bodies are preferable because they can search deeper and cover more ground. Both attributes will prove advantageous.

Focus your search on pockets ... and on float in dump sites in hillsides or in gullies, creek bottoms and other areas where you reason logical... that heavy pieces of float might have stopped. Once you begin to locate float ore, work upstream or uphill from your initial discovery and try to locate the source. Remember that ore is a combination material and will not produce as strong a reading as that from a solid metallic object. Do not overlook any faint signals. When you get a reading over a pile of rocks, lay down your detector and proceed to move these rocks, one at a time, across the bottom of your searchcoil as you practiced in your bench tests. By following this procedure, you can learn which rock(s) produced the signal.

Whether working in a creekbed or on a hillside, ground bal-ance your detector carefully and sweep the searchcoil two to six inches above the surface, depending on the "chatter" you get from iron mineralization. Listen carefully for faint signals that can come from pockets deep under an overburden or those that have only a low percentage of metal. (Some iron mineral may also be present.) Nevertheless, if the pocket is relatively rich conductivity and not too deep, you will most likely get a good (and positive) metallic response from your detector.

Chapter 4

Finding Gold

"Can I find gold with a metal detector?"
This is the question that we hear most often from gold seekers who doubt their ability – or anyone's ability, for that matter – to find gold electronically.

The answer to that question is a most emphatic, "YES!"

We have proved, many times over, the abilities of metal detectors for finding gold. Some of their results and those of others are listed below.

It is true that some manufacturers and promoters have misled many metal detectorists. Unfortunately, they have become convinced that they can rush out into a gold-producing area with *any* type of detector and find nuggets and placer gold. It is possible that those individuals who manufactured and/or sold the so-called **gold-finding** detectors simply did not understand the limitations of their instruments. Let us hope that naivete alone explains the situation!

The fact of the matter is that until almost the 1980's the Beat Frequency Oscillator **(BFO)** detector was the only type that could be depended upon to produce consistent and accurate readings. The BFO itself, however, was and continues to be limited because of its marginal depth penetration in mineralized ground.

Today's Very Low Frequency **(VLF)** ground cancelling (balancing) detectors offer greatly expanded capabilities:
- Detection depth has been tremendously increased, particularly with the highly regarded 15 kHz "Groundhog" circuit;
- Ground mineral problems have been mostly overcome;
- Rapid and accurate identification of "hot rocks" is possible, even for the beginner.

As a result of extensive field and laboratory tests and careful electronic design and manufacturing techniques, detectors possessing very exacting metal/mineral locating and identifying characteristics are now being built. Using computerized and other more modern versions of these tested and proven VLF detectors, both professional and recreational prospectors are making rich strikes in previously unworked areas, unearthing nuggets similar to those found at the turn of the century.

Significant recent gold and silver strikes include:

- In central Washington Roy Lagal (one of the authors) found a 2 1/2-pound gold nugget.
- A metal detector operator reports that he found a beauiful 36-ounce gold nugget near the central Cascade Mountains in Washington.
- The other author of this book, Charles Garrett, spent several weeks prospecting in Mexico; in one mine alone he found two large pockets and one vein of silver.
- A 47-ounce nugget found at Mt. Magnet, Western Australia, although valued at "only" $7,000 for its gold content, was appraised at $25,000 as a specimen.
- A 25-ounce nugget and a 36-ounce nugget were found with VLF detectors in northeastern Australian gold fields.
- Peter Bridge of Victoria Park, Western Australia, reports that more than 1000 pounds of gold nuggets have been found by his customers using VLF detectors.
- Two hundred ounces of small nuggets were recovered by two prospectors in Western Australia. One detector operator, working alone, found more than 500 ounces of large gold nuggets.
- After a gold panning weekend in Georgia, Pam Wendel Clayton returned home with a vial of beautiful, tiny gold nuggets.
- The most magnificent find of all may be the 62-pound Hand of Faith nugget found in Austrialia by an amateur prospector using a detector with the Groundhog circuit. This nugget was reportedly sold for one million dollars to the Gold Nugget Casino in Las Vegas, NV, where it is now on display.

Australian Gold

When a single man, equipped with only a VLF ground cancelling metal detector, finds more than 500 ounces of gold in a few short months, metal detectorists everywhere should take notice. A small-scale recreational mining boom has occurred in Australia, and recreational prospectors the world over have shown interest. Peter Bridge, who owns and operates a prospector's supply house in Western Australia, was largely responsible for introducing the VLF metal detector to that continent. Those who wish to learn more about his fascinating experiences and the gold situation generally in Australia should contact him at Hesperian Detectors, Box 317, Victoria Park 6100, Western Australia.

Although no "official" data is available concerning the overall results of the Australian boom, a newspaper in Perth reported, "It is estimated that more than 2,000 ounces of alluvial gold was picked up by Western Australian prospectors" in just the first two years.

This new Australian gold rush is attributed to two major factors:

– Banking (gold) regulations were repealed, enabling gold to be sold on the open market in Australia;

– Sophisticated VLF metal detectors were introduced by Peter Bridge and others to give Australian prospectors a highly efficient gold-finding tool.

In the State of Washington

Back in the United States we joined Frank Duval at Liberty, WA, to search the old dredge tailings, conveyor piles and mine pits for nuggets and rich ore deposits. Even though that area of the country is very highly mineralized, results produced by new ground balancing detectors proved truly amazing.

"Mining in this district has been described as different from any other mining district in the world," Roy Mayo writes in his book *Give Me Liberty.* "Placer gold has been found as water worn, rounded nuggets, indicating travel over a long distance. Nuggets have also been found having sharp edges with pieces of quartz still attached, indicating nearness to the source. Flakes of gold and delicate wire gold are also found. This is one of the few places in the world where gold may be found in its crystalline form."

Lode gold can also be found around Liberty. Shale and clay pockets contain delicate wire gold as well as gold in other forms in stringers in the sandstone and in a mixture of quartz and shale that is known as "birds-eye quartz." Major area discoveries of wire gold have been made on Flag Mountain and Snowshoe Ridge, according to Mayo. Many beautiful specimens of wire gold weighing several pounds have been found there. More information and this excellent book is available from Mayo Maps, P. O. Box 184, Enumclaw, WA 98022.

We have engaged in extensive electronic prospecting in the Liberty area. Because of its wide variety both in types of gold deposits and in types of hunting, it offers an unusual challenge. We worked old dredge tailings where large nuggets have been found by eyesight alone over the years by casual vacationers and picnickers. We also checked for pockets of wire gold on Mrs. Bertha Benson's claim on Flag Mountain and pin-pointed both non-magnetic conductive ore veins and magnetic non-conductive ore veins as well. All of these activities were coordinated by Jacob Kirsch, who has been mining claims in the Liberty area for many years.

Working dredge piles along Swauk Creek was a real eye-opener. After we first carefully regulated the ground balance of our detectors to cancel out existing ground mineralization, we swept the instruments slowly over the large boulders, searching for elusive gold nuggets that escaped the trommel of the dredge many years ago. A trommel is the sieve-like device on a dredge that sizes materials being dredged. Apparently this dredge had used a trommel with small holes because many nuggets larger than the openings in its screen were reclaimed by the creek as they passed over the screen and were diverted into the tailings piles.

This book has said little yet of "hot rocks," but it is a subject about which you will hear much more. On this pile of dredge tailings in Washington we really discovered the true meaning of the term, hot rocks. They can be defined basically as those out-of-place, highly mineralized rocks that try to fool an electronic prospector and his equipment. He can be in big trouble, indeed,

22

if he is not alert to them. As we scanned these tailings, we encountered countless numbers of hot rocks, but easily identified them with the calibrated discriminating capabilities of our modern detectors.

We worked a considerable area of those tailings, but the end results of the day's electronic prospecting were well worth the trouble.

On the hill above Liberty we carefully searched for pockets of gold and highly mineralized veins on a rich hardrock claim that has produced many thousands of dollars in gold. A number of veins were discovered and mapped out to be excavated and examined at a later date for precious metals. Initial findings proved conclusively that metal detectors with ground balance properly adjusted to cancel out mineralization can be of enormous benefit to any prospector. In fact, we are convinced that it is the most important tool that a weekend miner or recreational prospector can own. Used properly, the detector can find conductive (non-ferrous) and non-conductive (predominantly magnetic) deposits, as well as nuggets, pockets and veins of all kinds wherever sufficient quantities of magnetic iron are present.

Chapter 5

The Detector as a Tool

Reflect for a moment on the gold prospectors of the 19th and early 20th century . . . the hardships they endured . . . the difficult conditions they faced. Consider the total absence of any equipment to find and identify gold other than their eyes. If they could not actually see the gold itself or recognize gold-bearing rocks, they would not emerge from their struggle with nature bearing any color or nuggets.

On the other hand, today's prospector – whether he is working full-time at the task or only on weekends and vacations – carries the electronic metal detector as a valuable prospecting tool. Sensitive, lightweight, stable and reliable, today's professional metal detector serves as an "extra pair of eyes" for the modern prospector.

These detectors can find gold in a number of forms:
– Lode or hard rock deposits in a vein and often mixed with other materials;
– Placer deposits, either in a stream or in dry sands and gravel;
– Nuggets of pure metal, which can be found in nature by themselves, in veins or as part of a placer deposit.

Placer gold is the type most often sought after by the weekend prospector who enters the gold fields only with a pan or panning kit. More and more treasure hunters are taking detectors with them to search for nuggets and rich lode deposits as well.

Placer deposits are generally formed by weather erosion of an outcropping of lode gold. After the outcropping has become exposed over the years and decades, it begins to break apart and smaller pieces break off. Gravity and water run off carry them downhill. Some of these pieces of ore still lie on mountainsides awaiting a metal detector. Most of them, however, are

Gold Panning Instructions

1. Place the classifer atop the large gold pan and fill with sand and gravel shoveled from bedrock.

2. Submerge the classifier contents under water and use a firm, twisting motion to loosen material. Gold, sand and small gravel will pass through the classifier into your gold pan. Check for nuggets in the classifier and watch for mud or clay balls that might contain nuggets.

3. Discard all material remaining in classifier. With your hand thoroughly loosen all material in the gold pan. Inspect contents and remove pebbles. With the pan submerged twist it with a rotating motion to permit the heavy gold to sink to the bottom.

4. Keep the contents submerged. Continue the rotating, shaking motion. From time to time tip the pan forward to permit water to carry off lighter material.

5. As you shake the pan to agitate the contents, make certain that the Gravity Trap riffles are on the downside. The lighter material will float over the pan's edge while the riffles trap the gold. Rake off larger material.

5

6

7

8

9

10

6. Develop a method of agitation with which you are comfortable. Back and forth . . . round and round . . . or whatever suits you. Your aim is to settle and retain the heavy gold while letting the lighter material wash across the riffles and out of the pan. As its contents grow smaller, smooth and gentle motions are mandatory. Use extreme care in pouring lighter material over the side. Submerge pan often and tilt it backward to let water return all material to the bottom of the pan.

7. If there is a larger than usual concentration of black sand, you may wish to transfer the material to the smaller finishing pan for speedier separation.

8. Continue the panning motion to let all remaining lighter material flow off the edge.

9. Now, you can retain the black sand concentrates or continue gentle motions to let it ease over the edge of the pan. As visual identification becomes possible, a gentle swirling motion will leave your gold concentrated together.

10. Retrieve your gold. Use tweezers for the larger pieces and the suction bottle to vacuum fine gold from the small amount of water you permitted to remain in the pan. Save the remaining black sand for later milling and further classification at home.

carried by water down into streams where the large chunks are ground into sand and gravel, thus releasing gold from the lode. Because of its comparatively heavy weight, gold sinks and works its way vertically on down to bedrock of the stream where it is trapped in cracks or crevices while the lighter sand and gravel are swept on by the rushing waters.

While most placer gold was originally deposited by water, nature's changes over the centuries have left many good placer deposits high and dry. Sand washes in gold-producing country sometimes contain a high concentration of fine or so-called flour gold that can be recovered by dry panning or by use of a dry washer. It is not unusual to find, mixed in with placer gold, small nuggets that can be easily detected with the appropriate metal detector.

Working vein or lode gold can be more difficult. Generally, extensive experience or considerable research is required if the weekend miner is to be successful in developing new locations. Knowledge of geology is helpful, but the inexperienced prospector can often be quite successful simply by working around abandoned mines and mine dumps where others have extracted gold from nature. Many times, the early day miners would miss a rich ore vein by just inches and leave the location, considering it barren. Because they had only eyesight to guide them, these pioneer gold seekers would often carefully follow a vein far into a mountain, digging just inches away from an incredibly richer vein they could not see. Your modern metal detector when worked along the walls of these old mine tunnels may just reveal the precise location of that incredibly rich vein the earlier miners overlooked.

Today's weekend prospector finds all the hard work of moving tons and tons of earth already completed. In fact, by examining only the walls, roof and floor of a mine just a few inches deep, the recreational miner can prospect more cubic yardage in just a few hours than the old miners could in months.

While prospecting with metal detectors in Chihuahua, Mexico, we located three veins of silver that had been totally overlooked by early day miners. In scanning the walls of a mine owned by Javier Castellanos, Charles Garrett received three readings that demanded further investigation. Two of the readings proved to be silver pockets, and the third was a vein of native silver approximately one-half an inch in width. As Javier tooled his way into the mine wall, this newly discovered vein grew larger as it continued below floor level and proved to be several inches wide. This silver vein and the pockets perhaps would have been lost forever if the electronic metal detector had not signaled their presence.

Don't overlook the abundance of ore that is always to be found lying around on the tunnel floors or beside rails between the mine and mill or loading area. By testing these ore samples with a detector you can uncover valuable specimens overlooked by the original miners. No matter where the ore came from in a mine, when it fell off a cart, it was often not considered valuable enough to recover – because it contained no gold that was visible to the naked eye, the only gold-seeking tool possessed by the miner.

Similarly, dumps and tailing piles of old mines are good locations to work. They are certainly the easiest and quickest sources of gold if the prospector knows how to use a modern metal detector properly. Large gold or silver nuggets can be found concealed within a chunk of rock that was unknowingly discarded on the dump. The old-time miner could not see the valuable material hidden by the rock, but a metal detector can certainly signal its presence.

In Cobalt, Ontario, Canada, we located several large chunks of silver that were apparently discarded accidentally 50 years ago along with worthless material that was used to build a roadbed. The largest piece of ore we detected weighed 50 pounds and consisted of 85 percent silver.

Never pass up the opportunity to work discards in the tailing piles around abandoned mines. Many times more gold is still in the ground that was recovered by the early miners. It is yours for the taking!

Chapter 6

The Pan–Detector's Companion

Τ he modern metal detector and a good plastic gold pan can prove to be a dynamic duo indeed for the weekend prospector. Let your metal detector locate nugget deposits and good panning locations. If your search is for nuggets in a stream bed or creek containing rocks with high mineral content, a sensitive, ground balancing detector is invaluable. In fact, it will probably be impossible to detect nuggets in such a location without this instrument. When you have precisely adjusted the detector's ground balance to eliminate the highly mineralized sands, the instrument will be super-sensitive to any conductive gold nuggets, even the tiniest ones.

When engaged in electronic prospecting, a plastic gold pan often facilitates target recovery. Many times it can prevent arduous digging or abandoning a target that has fallen down into a large pile of gravel or large rocks. Whether this target in a rock pile or flowing stream proves to be only a spent bullet or a valuable nugget, the plastic gold pan can make its recovery much faster and easier.

The technique is simple. First, of course, pay close attention to the faintest signals from your detector. They may indicate that the nuggets are small or that they are deeply buried. Second, you must pinpoint each detector response as closely as you can. Then, slip a shovel well under the spot from which the signal came. Be extremely careful now since small objects made of heavy metal always will sink into gravel and may become lost. Place into your plastic gold pan all of the gravel or sand that you scooped up, and scan this sample with your detector.

The value of having a *plastic* pan is now apparent. It would be impossible to scan the sample if the pan were made of metal. In the plastic pan your conductive target will respond. If there is no response, dump the pan and scan the location again until your detector signals. Pinpoint your target and try to get under it again with your shovel. With a bit of practice, you will be surprised at how quickly you can become proficient in this technique – even when working in three feet of water!

When using this technique to simplify recovery and save time, you will find that use of a pan with the built-in Gravity Trap principle provides not only a quick and efficient method of sorting through sand and gravel to locate the metallic object that caused your detector to respond, but that also lets you quickly separate ALL the heavier concentrates from lighter material and facilitates panning that may be necessary.

Your techniques of recovery using a plastic pan are the same when searching a dry wash or placer diggings, except the object will be easier to locate than in a stream. Old dredge tailings may be somewhat more difficult in which to locate targets, and you may lose a few here before you master the technique. Material in which you are working is loose, and an object of heavier metal can easily and quickly work its way deeper into the pile of tailings. Once lost, they are often very difficult or impossible to recover since they simply work their way deeper with each attempt you make to recover them.

You will be tempted to give up and move to a new location. On the other hand, you'll be surprised at how quickly you become proficient and tenacious at such recoveries once you realize that the metallic object you are seeking just might be a gold nugget!

Chapter 7

Metal/Mineral Identification

Two factors generally determine whether non-ferrous metals will produce a metallic signal when encountered by a metal detector – the quantity of metal that is present and the physical state in which it is found.

When the electromagnetic field of a metal detector's search-coil is disturbed by a sufficient quantity of gold, silver, copper or other valued non-ferrous metals, the detector signals a metallic response – provided the metals are in a conductive state. Since some extremely rich ores are in sulfides, tellurides and other compounds that are not conductive, they do not produce a metallic response, regardless of their quantity or purity. Free milling ores of non-ferrous metals, however, generally produce good responses when the ores are encountered in sufficient quantities by a metal detector.

About the only mineral that a metal detector can recognize as **mineral** is magnetic black sand or magnetic iron, Fe_3O_4. A detector tuned in a true calibrated test mode can easily determine whether the ore contains a predominance either of metal or of mineral. If the ore specimen contains neither metal nor mineral or absolutely equal amounts of both, the detector will produce no indication.

A *mineral* response from the detector does not necessarily indicate there is no metal present; rather, that there is a predominance of mineral. When the specimen signals a metallic response, you can be certain that it contains metal in conductive form in such quantities that you should investigate the specimen thoroughly. Such response capabilities of modern detectors with

37

discrimination make it today's most important field tool for the identification of metal vs. mineral.

Make Your Own Samples

To understand how a metal detector signals the presence of metal, you should make your own "ore" samples. A U.S. copper penny provides one of the metal samples you will need.

Producing your mineral sample will require a little more effort. Place a large iron nail or a piece of soft iron into a vise and file the nail with a very fine file. Place a piece of paper under the vise to collect the filings. The amount of filings required is about equal to the weight of a silver dime. Place these iron filings into a small plastic container (a medicine pill bottle is satisfactory) with a diameter about equal to that of a dime. Fill the bottle with glue and let it solidify. You have now produced a sample of non-conductive iron mineral that will cause a response from your metal detector identical to that which is caused by much of the iron mineral you find while prospecting. Remember that you will need a very fine-toothed file to make your filings of almost powder consistency!

Your next sample will demonstrate the difficulty of detecting silver oxides, gold dust and wire gold. Reduce a penny completely to filings. Place these particles in the same size bottle as your iron filings and again fill with glue to hold them permanently. You now have what is basically an ore sample. It is composed of marginally non-conductive, non-ferrous ore whose presence in the electromagnetic field of your metal detector will cause the instrument to produce a "questionable" positive response.

Scan these samples with your detector. Study carefully the responses generated by each. Such practice will greatly aid you to analyze veins and pockets when you encounter them in the field, and you will begin to learn how to identify the metal/mineral content of ore samples correctly. Try to obtain samples as many of the other ores mentioned in this book as possible.

Chapter 8

Prospecting With Modern Metal Detectors

Observe three basic rules for prospecting with a metal detector. Your success is not guaranteed; nor, will it be instant. If you follow these three rules, however, and are persistent, you can locate and identify any precious metals you encounter:

1. Select the *proper type* of detector. We refer, of course, to the operating characteristics of the detector rather than the brand name. We, obviously, recommend the new Garrett A3B-United States Gold Hunter, with its legendary Groundhog circuit that has found tens of thousands of ounces of gold in Australia and around the world, or the new Grand Master Hunter with its microprocessor-controlled, computerized circuitry pre-programmed at the factory. These two detectors are capable of locating even pin head-size nuggets and will also identify hot rocks properly.

2. You must learn *patience*. Understand your detector and its capabilities fully. Also learn its limitations so that you can become proficent in its use. Read this book carefully. Also read the Owner's Manual supplied with your detector. After you have practiced with your detector, read both books again and study them. To be successful with a metal detector it is essential that you understand it thoroughly as well as all of the techniques for using it that we describe.

3. Begin your search where gold is *known* to exist. The preceding sentence is one of the most important in this book! It is

impossible to find gold or any other precious metals where they simply do not exist. Prospectors who have found gold with metal detectors have demonstrated wisdom and patience, plus a great deal of **research.** Stick to known, productive mining areas until you are familiar with your detector's operation and understand how it signals the presence of mineral zones. Soon you will discover many other ways in which a quality metal detector can open up completely new areas of success for the prospector and recreational miner.

The Right Detector

To achieve the greatest success the ideal metal detector for prospecting is a sensitive, deep-seeking instrument that has been designed for prospecting and proved in field use. It must offer precise (adjustable in the field) ground balance and properly calibrated discrimination circuitry. Such a detector must be calibrated correctly by the manufacturer for metal vs. mineral determination and hot rock identification. With power to penetrate the toughest iron mineralization, it should have the availability of a complete range of searchcoils from about 4 1/2 inches in diameter to more than 12 inches.

Garrett's A3B Gold Hunter offers 10-turn controls for both ground balance and audio threshold. These permit the extremely precise adjustment demanded in heavily-mineralized areas. Nearly identical ground balancing and detection characteristics are offered on Garrett's Grand Master Hunter and Master Hunter 5 detectors, both of which are ideal for prospecting.

It is possible, to some extent, to use *any* ground balancing detector to find precious metals. Most such detectors can now be operated even by a beginner in areas where the older models would have proved almost useless in the hands of an expert.

Automated – or, *motion* detectors, as they are sometimes called – can be used for nugget hunting. When using such an instrument, however, you must resign yourself to overlooking tiny nuggets. And, many of the motion detector's discrimination controls can not be properly set at the factory for metal/mineral discrimination. Thus, even when the discrimination controls are set at Zero (all metal), the detector still provides considerable discrimination that might cause gold to be lost.

Depending on existing mineral conditions, you might achieve some success, but we can almost guarantee that you will overlook nuggets, ore veins and the like that would be found with the *right type* detector. This instrument with properly calibrated ground balance and discrimination that has been proved by successful testing and use in the field will offer superior performance in detecting small and large nuggets and both conductive and non-conductive ore (predominantly magnetic) veins.

Correct identification of metal, mineral and marginal ore samples is critical, and it must be accomplished with the detector in a stable position. Otherwise, you and your detector are guessing. Detectors that discriminate or reject targets by operator manipulation, such as "whipping" or operating at a prescribed speed, are virtually useless in prospecting. Under almost 100% of any circumstances your *sample* must be moved across the face of a searchcoil in a controlled manner to assure correct identification. You should also remember that you will often be operating in areas (mine tunnels, for example) that are too cramped to permit whipping. Basically, coin hunting detectors that discriminate by *ignoring* the target or by requiring operation in some unusual manner should not be used in prospecting. You are probably wasting both time and opportunities if you take them into the gold fields.

During the early days of metal detecting, much literature was produced (by these authors, too!) on prospecting with detectors. The majority of that writing revolved around the ancient, but trusty, workhorse of the industry, the beat frequency oscillator (BFO) metal detector. Both of us can become almost romantic in recounting our successes with it. We loved that old detector, but the simple fact is that today the BFO is obsolete. It is no longer manufactured. Anyone who has been successful in the gold fields with a BFO has found this success magnified MANY times over through use of proper modern detectors. Our experiences conclusively prove the truth of that statement.

Searchcoils

Two types of searchcoils are commonly available for use with prospecting instruments, the co-planar loop and the concentric loop. Both searchcoil windings are positioned on the same plane, but only concentric windings are centered around the same vertical axis. Consequently, concentric searchcoils produce a desirable, uniformly distributed electromagnetic field. Fewer irregularities will present themselves when a concentric searchcoil is used.

Bench Testing

1. Odd-shaped samples should always be held in several different positions when testing. An elongated (football-shape) sample, for example, could give one reading if held broadside and a different reading if the pointed end faces the coil. Flat metallic samples produce larger detector signals when the flat side is parallel to the bottom of the searchcoil. On the other hand, iron mineral samples produce approximately the same negative signal, regardless of their orientation.

2. Always move the sample across the searchcoil. Reasonably rapid motions should produce speaker variations that can be heard easily.

Adjustments for Sampling

Metal/mineral ore sample identification can be achieved in the Discriminate mode if the manufacturer has designed circuitry with properly calibrated adjustments for discrimination controls.

Discriminate Mode: Rotate your detector's discrimination control knob to its Zero setting, usually fully counterclockwise. At this setting your detector should respond positively when *any* metallic object is brought across the bottom of its searchcoil. Even tiny nails, regardless of their orientation, should produce a positive speaker sound when brought across the searchcoil. Remember, **all** metal objects **must** produce a positive indication. If this cannot be achieved, the discrimination control has not been correctly calibrated.

We have illustrated our comments concerning phases of detector use with Garrett models simply because we are most familiar with them. Other manufacturers produce good detectors, and you can apply our instructions to them provided these detectors offer stable ground balance and precise calibrated discrimination.

To sum up – if your detector has been correctly calibrated (**every** detector manufactured by Garrett is tested before it leaves the factory), you can check ore samples accurately in the Discriminate mode simply by setting your discrimination control to its lowest setting.

Analyzing Ore Samples

Before you take your detector into the field, experiment with it on the bench to learn how it will react to various ore samples. Use the samples you have already made and obtain additional samples of galena (lead), silver and gold ore, iron pyrite (fool's gold) and ordinary rocks. If possible, get two grades of ore for your tests, a very high grade and a very low grade specimen of each. By conducting your own bench analysis you can become familiar with the type and amount of detector response to the various materials.

If the ore you are checking has a predominance of metal in detectable form (use your solid dime sample for testing), you will hear a sound increase as the sample is moved across the searchcoil. If your detector is equipped with a sensitivity meter, you will see a positive pointer movement indicating the presence of metal. On the other hand, if the sound has a tendency to "die" slightly, your sample contains a predominance of mineral or natural magnetic iron (Fe_3O_4). As mentioned previously, this does not mean that the sample contains **no** metal, only that it contains **more** mineral than it does metal. If the sample contains **neither** metal nor mineral, or if it has electrically **equal** amounts of both, you will receive **no** response on your detector.

Always remember that metal detectors do *not* respond to many types of ores. Only those containing metal in conductive form in sufficient quantity to disturb the electromagnetic field will cause the detector to respond positively. Continually practice bench testing to become more familiar with the type and amount of response your particular detector gives to both low and high grade ores. A little time experimenting can save you many hours of confusion when you are in the field – and it may keep you from missing valuable nuggets, ore samples or veins!

Hot Rocks

A hot rock is simply a mineralized pebble that contains a quantity and/or density of low conductive iron mineral that responds as METAL when the ground balance controls of your detector

43

are adjusted to a particular setting. Always remember that what responds as a "hot rock" in one area will not necessarily give the same response in another area where you have set your ground balance controls differently.

To use a cliche and to use it aptly, these mineralized pebbles are the absolute bane of a prospector's existence. There is no getting away from them if you are diligent and persistent in your search for gold. Furthermore, they can drive you crazy if you let them. Let us emphasize how vital it is that you learn to understand these little pests and to know how to respond to them properly.

When you are operating with your detector in the All Metal mode, the signal you receive from a hot rock will be the same as you would receive from a metallic object. The signal is positive and unmistakably metal. Of course, you should first try to identify any target before digging. The method used to identify hot rocks is explained fully in Chapter 10. We mention the subject here because recognizing and identifying them is so important to an electronic prospector.

Hot rocks can be found in jumbled rock piles as well as in ground areas that look as if they have never been disturbed. Their iron mineral content, however, is so exceedingly different from that of surrounding rocks that they disturb the delicate ground balance of your detector to such an extent that they present themselves in its electromagneric field as metal targets.

Hot rocks are "freaks" or "oddballs" of nature. They should not be where they are, and they should not cause your detector to react the way that it does. But, they do! Simply stated, however, hot rocks are "there," and we must deal with them. With a properly calibrated detector you can identify them quickly, and they will become to you exactly what they are to veteran prospectors – tolerable pests.

Grid Patterns

Searching in grid patterns is a method that has proved successful in field prospecting when looking for veins. It offers the opportunity of approaching a vein or other type deposit from two different directions, avoiding the distinct possibility of walking parallel to a vein but never actually crossing it. Set up a fixed grid or crisscross pattern and sweep your searchcoil across it in wide, even strokes.

Always adjust your detector in the All Metal mode, balanced to the mineralization of the surrounding ground. Use manual audio tuning or the factory-set ultra slow auto audio retuning mode. Walk slowly in as straight a line as topography permits, scanning a wide path with the searchcoil ahead of you. When you reach the end of one of your "lines," turn and walk a parallel path in the opposite direction approximately 10 feet from the first path. Continue until you have covered the area selected. Then, repeat this procedure, except walk parallel paths at 90-degree angles to the first set of paths. Now, you have completed your fixed grid or crisscross pattern search of the area.

If your detector speaker sound increases and/or the meter pointer shows any increase while you are scanning, notice the intensity and the duration of the increase. You may have discovered a vein or ore pocket beneath you.

On the other hand, your detector may have changed its tuning due to atmospheric conditions, interference, bumping of controls or some other reason. Do not touch your detector's controls. Rather, return to that point where you were scanning before you noticed an increase. If the speaker and/or meter decreases to the previous level, your detector's positive response was caused by conditions in the ground, not some detector or operator problem.

Continue to retrace your steps. As you reach the point where the sound changed earlier, it should again change if you are detecting an ore vein or pocket. As you continue walking, pay close attention to the detector's audio responses. They should either increase further or drop off to your initial tuning level. If responses are "increasing," your ore is getting RICHER. When the responses return to your initial tuning level, you have walked on over or past the ore deposit. Plot or map the deposits or veins. If there are several, note where they cross or crisscross each other beneath the surface.

After you have found detectable ore deposits, identify their nature by operating your detector in the Discriminate mode. In this mode the detector will indicate whether your deposit is predominantly iron (or, low-conductive pyrites) or predominantly non-ferrous material. Veins crisscross each other beneath the surface, and a vein of gold may be bisected by several iron veins. Location of these veins can be plotted fairly accurately by paying careful attention to the responses from your detector. Depending on your ground balance setting, you may also receive positive metallic indications or negative indications from heavy concentrates (pockets) of magnetic black sand. Since such pockets often contain gold, they should always be investigated.

More information on the subjects of veins and ore deposits and the mapping of them is contained in Chapter 11.

Facing
Authors examine a gold nugget Roy Lagal found in this dredge pile. Note pan into which he shoveled rocks and sand to scan and look for detected targets.

Over   Top
Field testing of all equipment is important in detector development. Here, the authors use several instruments over highly mineralized ground.

Bottom
After Roy Lagal pans silver ore from a Canadian stream, he transfers contents into a smaller finishing pan for separation of silver and other elements.

A modern metal detector lets an electronic prospector check more area in hours today than could have inspected throughout the entire life of a mine.

Charles Garrett uses a 12-inch searchcoil to look behind the walls of an abandoned mine, seeking ore and veins that early-day miners overlooked.

Chapter 9

Nugget Hunting

his term is so ambiguous that no description of it could ever be complete. Much has been written about this method of searching for gold, generally with instructions that confuse the recreational prospector. We will concern ourselves simply with the basic instructions for finding nuggets in streams of moving water and in dry washes and old diggings.

Facing

Field searching with a modern metal detector can be profitable as the electronic prospector seeks pieces of float on the surface.

Over

After his Gold Hunter detector has located a target in a stream, Charles Garrett rakes material into a plastic gold pan for closer examination of the find.

"Working" a stream properly requires careful inspection behind large rocks where placer gold and nuggets could have been washed.

Searching for nuggets in a beautiful (and cold!) gold country stream, Charles Garrett has a plastic gold pan ready for examination of his discoveries.

Some manufacturers provide smaller searchcoils (Garrett's famed "Super Sniper" is 4 1/2 inches in diameter), but careful field testing by the authors has shown that regular searchcoils such as Garrett's 8 1/2-inch Crossfire are very efficient for nugget hunting, even for searching out the smallest bits of gold.

In Water

First, make certain that you are using a *submersible* searchcoil.

Carefully adjust the ground balance controls of your detector to eliminate interference from the highly mineralized black sands found in most gold-bearing stream beds. Use earphones to heighten your awareness of even the faintest detector signals. Earphones will also help prevent the sometimes loud noises of running water from masking detector signals.

Operate the searchcoil from about four inches or higher above the bottom of the stream, moving it slowly over your search area. Operating height will depend upon mineralization of the area and the amount of "chatter" these minerals are causing you to hear.

Two valuable tools will improve your ability to recover gold nuggets in a stream:
 – A plastic gold pan;
 – A trowel or shovel, depending on the depth of the water, and a small pry bar to loosen compacted rocks and gravel.

When you discover a metallic target, remove a few inches of sand and gravel and place it in your plastic gold pan. Test the entire pan of material with your detector to determine if you have recovered the target that produced the original signal. If the target is not in the pan, dump its contents back into the water, locate your target again with the detector and scoop up another shovel of material from the stream bed into your pan. Continue repeating this process until you locate your target in the pan. Even then, recheck the hole for additional targets.

When you are certain that you have the metallic target in your gold pan, first try to locate it visually by sorting the rocks and pieces of gravel carefully. If you cannot spot it, concentrate the material by panning. If you are still unable to find the target, perhaps it is only a small piece of ferrous trash like a boot nail. It could also be a highly mineralized hot rock with mineral content different from than that which your detector is adjusted to eliminate. Use the regular procedures for checking for hot rocks.

This same method of detection can be used on old dredge tailings either in the water or on banks of streams. Practically all rock piles present on backs alongside streams are dredge tailings. Old dredge tailings have produced some fantastic gold finds for treasure hunters with metal detectors. Never overlook these opportunities!

Dry Washes and Diggings

Old placer diggings and the bottoms of dry washes are often productive locations for the discovery of nuggets. Knowledgeable nugget hunters have profited over the years by working dry or desert areas in highly mineralized locations and by working areas where detectable gold nuggets were found by early day prospectors. In these remote desert areas where water has never been available to prospectors and the only method of recovery is with a dry washer or by dry panning, millions of dollars worth of small nuggets await the recreational prospector with a metal detector. These nuggets are rarely detectable by eyesight, but they generally lie on the surface or at very shallow depths. Investigating low areas with a properly ground balanced, sensitive detector can be rewarding. A modern metal detector is the only practical method for locating these small nuggets. It is certainly the quickest!

Your modern ground balanced metal detector is the best choice for nugget hunting. Searchcoils about 8 inches in diameter will search the greatest depths, yet remain sensitive even to the most tiny nuggets. Precise ground balancing is necessary to overcome the mineralized conditions of most of these prospecting areas.

After you have correctly ground balanced your detector according to manufacturer's instructions, scan with the searchcoil held about an inch or two above ground surface. **Caution:** Searchcoil operating height will be determined by the amount of mineralization present and the size of rocks lying on the surface. You do not want to listen to any "chatter" from the jumbled mess of mineralized material; especially, since it might cause you to overlook the true signals of conductive metals that your detector will transmit to you. Operating heights of greater than two inches may be required in highly mineralized ground.

Always remain confident that your modern detector is doing a good job for you. It will penetrate iron mineral soil to detect nuggets. Remember, however, that *all* detectors are not suitable for prospecting and that some hobbyists have actually given up this interesting and rewarding pursuit because of poor results that were wholly attributable to their choice of detectors.

Physical recovery of nuggets from dry sand can be accomplished in the same manner you retrieved your target from a stream. The process should be somewhat easier without water rushing around you and your feet slipping on wet rocks! Just slip your shovel in the sand where you get a signal. Place the material in a plastic gold pan and check it out with your metal detector. If you fail to get a reading, dump the pan and repeat the process. When your detector verifies the target is in the pan, make a visual search. If you are unable to locate the nugget, dry pan until you find it.

A good detector with precise ground balancing and calibrated discrimination will amaze you with the number of things it enables you to recover from the desert. Despite the advertising claims that you will read, there are only a few such detectors. We can speak from experience about the capabilities of Garrett's A3B-United States Gold Hunter, Grand Master Hunter and Master Hunter 5. Detector practice and experience in gold nugget country could very well lead to success. Gold in your pouch will make you feel like an expert . . . which, of course, you can become!

Chapter 10

Searching Old Mines

Wen you find an old mine that you consider a likely source of gold, begin by checking its floor with your metal detector. There is always the possibil-· ity that you are the first person who has looked, and you can often find high grade ore in older mine tunnels and shafts. Remember that they were worked by less modern methods than you are using.

Few weekend miners seem to realize that tunnel floors of old mines can be some of the most productive areas for searching with a modern metal detector. *All* of the ore that left that mine crossed the tunnel floor at least one time. Some pieces of valuable ore almost certainly fell from carts. It is certain that all fallen ore was not recovered but eventually became covered with rock and other debris. This high grade ore has lain there unnoticed for years, just waiting for some lucky treasure hunter to find it.

Begin your mine search by ground balancing your metal detector to cancel out ground mineralization. Make a general sweep of the tunnel floor, collecting all samples that give a metal reading. Remember, though, that there are likely to be many small iron objects in an old mine – rail spikes, pieces of the rail itself and hangers for candles and mining lamps. Spikes that were sometimes used to secure the tunnel shoring will still be in place and can be picked up by your metal detector. If rails are still in place, you can work closer to them by detuning the audio and using smaller searchcoils.

Once you have collected a number of likely looking ore samples, lay your metal detector flat as you did for bench testing. Test each of your rock samples individually. Continue this testing as you work the tunnel thoroughly from one end to the other. Use your rock pick to dig beneath the surface debris and test all small, likely looking samples.

Here's an additional tip on this type of mine searching: The "pay streak" will normally always contain a much higher metal or mineral content than the ordinary rock of the tunnel. For this reason save *any* unusual samples for later careful examination.

Mine Pockets and Veins

Many of the old miners missed pockets and veins of high grade ore. Because of the limitations of their equipment, they passed within inches of the gold and silver that they sought as they chewed into the mountainside in their search for new deposits. Modern detectors can easily pinpoint these valuable pockets and veins for the electronic prospector.

Since you will find most mine tunnels driven through highly mineralized or magnetic material, your modern detector must be ground balanced as precisely as possible to cancel the effects of these minerals. We also recommend that you always tune your detector so that you maintain a faint but constant sound from the speaker. (Of course, headphones are always better!) Operate your searchcoil approximately four to twelve inches from the tunnel wall, depending upon the amount of iron mineralization present. Scan the walls and ceiling carefully, marking or taking note of any postive (metallic) signals. We find that a can of spray paint makes a good marking device. Ore containing a sufficient amount of conductivity (and some low-conductive magnetic ore "hot rocks") will respond positively as metal.

How to Identify Hot Rocks

Ground balancing your detector properly to compensate for a high mineral background will cause it to give a metal response to hot rocks (and hot spots). This occurs because the hot rocks and hot spots are simply isolated or out-of-place minerals for which your detector has not been ground balanced. (See Chapter 8). Identification of hot rocks and hot spots is no problem whatsoever to a modern, properly calibrated detector.

We reemphasize the importance of both proper design and proper calibration by its manufacturer for universal use, including prospecting. ALL metal detectors are *not* suitable for prospecting, despite claims that are made for them.

The procedure for identifying mineral hot spots (and hot

rocks) is a simple one, but it will require practice from you. To check to see if the "metal" response is metal or mineral pinpoint your target with the detector in the All Metal mode. Then, switch to the Discriminate mode of operation. The discriminate controls should be set to zero or to the level specified by the manufacturer as the calibrated level for ore sampling. Audio retune the detector, if necessary, according to manufacturer's instructions. (Computerized Garrett detectors retune automatically.) Now, with a constant sound (threshold) coming from your detector, pass the searchcoil back over the target. If the sound level decreases (or goes silent), your target is magnetic iron ore or oxides. These are the *only* substances that will cause the signal to stop. When this happens, ignore the target, switch back into All Metal and continue searching.

You have just exposed a *hot spot!*

If, on the other hand, your signal increases or remains steady, the target should be investigated. Increase your variable discrimination control (usually by turning the knob clockwise) to determine the amount of conductivity in this target. This procedure makes it obvious why you should not attempt to prospect with a detector that has a fixed discrimination control. If you have previously practiced with varying discrimination, you already know the approximate point on the control where worthless pyrite will be rejected. If you continue to receive a positive response after you have passed this setting, it is very possible that you have discovered a non-ferrous pocket or a vein of conductive ore.

You may have just *struck it rich!*

Remember that iron pyrites have a much *lower* (poorer) conductivity factor than non-ferrous ore. Pyrites will be rejected by the discriminate circuitry while the detector still accepts low grade non-ferrous ores. Keep in mind that all ore bodies may not necessarily be solid metal, and your response may be rather faint unless the ore is of extremely high conductivity. Remember your homemade non-ferrous ore sample, and don't overlook weak signals. Detecting non-ferrous, low-conductive ore can be very difficult. Of course, the ore content must still be determined by analysis or assay, but this important detection capability of prospecting instruments is largely overlooked, even by professional miners and geologists.

Elsewhere in this book we describe the method by which you can analyze detectable substances using your Discriminate mode. Remember, that while conductive analysis is possible in the Discriminate mode, only chemical assays produce qualitive and quantitative data.

To those of you who have worked in mines with BFO detectors, we urge that you return with a modern ground balanced instrument – especially, if you considered your search successful. Those "negative" BFO readings you ignored may contain much more than magnetic iron!

A Word of Caution

Always remember these warnings:
– Abandoned mines can be very dangerous.
– Never work alone around a mine.

When you are exploring or working around deserted mine shafts and tunnels, extreme caution should always be exercised. Shoring timbers have rotted over the years, and water seepage may have loosened once-solid tunnel walls. Any loud noise or impact against the timbers or walls of a tunnel could bring a mountain down on the old mine.

Equal care should be used any time you even peer down an old mine shaft. Not only is there a chance the earth at its edge could crumble, causing you to loose your footing, but poisonous fumes coming from old mine shafts have been known to kill people. And, they weren't even prospectors, just tourists who wanted to look down an old mine!

Never let all members of your party enter a mine at the same time. Someone should always remain at the surface or at the entrance to a mine to summon help, if needed.

Don't forget the laws of ownership. Just because an old mine *looks* deserted, you do not necessarily have the right to enter, much less begin searching with a metal detector. There is probably somebody with a valid claim to that mine, and he or she might object to your doing a little high grading on their property. Some old timers object violently . . . and they show it! It is always best and proper to gain permission to search. Very few claim holders will object to your electronic prospecting, particularly if you offer to give them helpful information on any veins or mineral pockets you might locate.

Old Mine Dumps

As you travel through the deserted gold fields, you will observe many abandoned tunnels, rock heaps (tailings) and piles of ore that never made it to the stamp mill to be crushed. By using your metal detector properly to test these rocks, you can often recover many good ore samples.

When you search for high grade ore in a mine dump, remember that during the working life of the mine, a tremendous amount of rocks and other debris from shafts and tunnels may have been deposited in that dump. Only a small portion of it may consist of tailings from the vein itself. Thus, you can readily see how important it is to take rock and ore samples from many different locations, especially the higher sections, of the dump.

To check for high grade ore samples lay your detector flat and tune it precisely (just as you did for bench testing). Then, move your selected samples, one at a time, quickly across your searchcoil to test for metal. After a reasonable number of samples from one location have been tested and no metal has been located, move to another area of the dump. If, after prolonged testing, you do not recover at least a few metallic specimens, it may be that the ore was of very low grade composition and is not of a conductive nature.

Do not overlook the possibilities of working these old ore dumps. Old time miners worked hard for hundreds of hours to bring these rocks from beneath the earth. Many beautiful silver and gold specimens of considerable value have been found this way . . . found hiding inside worthless-looking chunks of mine tailing debris.

You can scan over the rocks, but in most cases it is very difficult to correctly ground balance a detector. Since each rock may cointain a different amount of iron, just a slight movement of the searchcoil can result in a significant change in the minerals beneath it. This can result in rapid and pronounced ground unbalance of the detector. Also, there may be many "hot rocks" present in the dump.

We urge you to try it anyway! Choose a spot where you are

able to ground balance your detector adequately, then scan several inches above the rocks rather than close to the surface.

Chapter 11

Field Prospecting

The title of this chapter, or **field searching** as it is sometimes called, could cover the entire spectrum of recreational prospecting. It encompasses a wide variety of searching for precious metal, such as locating deep veins, looking for placer or nugget deposits, hunting for rich float material and pocket hunting.

Remember that volumes have been written concerning most major geographic areas where large discoveries have been made. We urge you to conduct research and to make use of as much of this material as you can. Furthermore, you must search for gold in areas where it has been found. Technical manuals detailing the discoveries, however, are written by mining engineers, professional prospectors and others in the mining field, many of whom – despite their expertise – are not familiar with modern metal detectors. Our suggestions in this book may seem quite simple, yet they pertain specifically to electronic prospecting with a modern metal detector and include procedures used by most successful electronic prospectors today.

Field prospecting is where all the knowledge you gained previously in your bench tests will pay off in big dividends.

Ore Veins

Contrary to popular belief, large mining companies do not conduct nearly the amount of field prospecting that many amateur prospectors imagine. Profit is the reason. A large corporation spends its time in efforts that directly result in profits. Since prospecting takes a great deal of time with no guarantee of profit, the big companies rely to a large extent on individual prospectors to make finds that they can exploit.

61

If you should be fortunate enough to locate a rich ore vein or deposit, it can often be sold to one of the large mining corporations. The recreational miner who enjoys just being in the wilderness for a relaxing weekend or vacation, whether gold is found or not, is more apt to seek out the difficult gold deposits than are the large companies. Searching for gold will take you into some of the most beautiful areas of our great country. It will offer you the opportunity to camp in fantastic natural settings. Yet, your efforts still might result in that once-in-a-lifetime chance for "the really big one."

All types of gold deposits originate as an ore vein formed during the volcanic activity of past eons, and such veins may go deep into a mountain and be fabulously rich. When such a lode is located, mining companies will literally jump at the opportunity to purchase rights to extract ore from this vein and smelt it into pure metal.

Deep Veins

Deep veins are usually a composition of several metals and minerals. For this reason extreme vigilence must be exercised because signals from a deep vein may sound very faint even through your headphones and especially from your detector's speaker.

A vein may be either metallic and respond as metal, or it may have a predominance of iron oxides. Evaluate the signal by considering its magnitude, the direction in which the vein appears to lie and how far it may possibly run.

In seeking deep veins use the largest coil with which your detector is equipped. Pay close attention to all responses. Investigation of irregular or unusual signals will many times lead you to pay dirt.

Pockets and Veins

Searching for and finding surface **float** can narrow your search area. Use medium-size searchcoils when searching for float. Then, when you think you have found the location of a vein, switch back to your largest coil and work in a grid pattern. Set the detector's tuning to enable you to hear continuously a slightly audible threshold sound and make wide sweeps with your searchcoil. Because the response area may be quite wide, you should cover enough area in your sweep to be able to determine the edges (start and finish) of the signal. Within reason, the amount of response from your detector's signal should permit you to judge the depth of the vein.

Mapping Vein/Ore Deposits

Once you have detected an ore deposit, you will want to know whether you have discovered an isolated pocket only a few feet in diameter or an intricate network of iron and non-ferrous veins that cross and crisscross each other. You will find that mapping your newly found deposit is both challenging and intriguing.

One reminder before you start mapping: make certain your detector is precisely ground balanced for mineralization in the area you seek to map.

Follow the grid pattern searching techniques described in Chapter 8, paying careful attention to detector responses as you plot them on paper. Responses by your detector from the veins below will vary widely, according to the width of the veins and/ or how the iron oxides have leached out into the surrounding earth matrix. It is possible that a vein only one foot wide can leach out such an area that it appears several feet wide to your detector.

You may find that your detector response will increase and remain at this greater level for some scan distance before returning to your present tuning. On the other hand, detector responses may change quite often, indicating the possibility of a network of veins. Carefully evaluate those areas you are scanning when your detector's responses change several times. While standing over a vein or deposit, check both the ground balance of your detector and its threshold level. Pay close attention to the detector's responses as you walk atop the vein or deposit. Try to determine not just if the detector responses change but also how much they change. Variations can indicate several magnetic ore pockets or perhaps contact points where veins bisect ore deposits or other veins.

In areas that have previously been worked by prospectors, careful readings can many times tell an experienced miner of an extremely rich pocket or the place where a rich vein has "slipped" and become "lost" from known pockets and veins. The seasoned prospector will generally be able to correctly identify all deposits found by a detector. Of course, core drilling to obtain samples for assaying is the easiest (and, sometimes, the only) way to determine the ultimate value of a discovery.

Placer Deposits

Placer mining is probably the most popular type of recreational prospecting. It offers the amateur miner a greater chance of recovering gold at lesser expense. This is because less equipment is required than in other types of prospecting. In addition, such vicarious pleasures as the enjoyment of nature and healthy outdoor exercise seldom let placer mining become a chore. It is always a pleasure!

Even as you read this, your chances for placer mining success are getting better as rock deposits in gold country are eroded by wind, water and normal earth movements. These actions of nature release precious gold from the veritable prison of rock in which it has been held captive for centuries. As the heavier gold seeks lower levels, run-off from rain and melting snow moves it down mountain slopes into freshets and brooks which carry it into larger streams. During periods of water movement, streams in the gold fields are heavy with dirt, rock, other debris and gold. As gravity carries everything along, heavier materials – which includes gold – become trapped along the way as they sink below the lighter dirt and rocks which are being swept downstream.

Nature offers placer gold countless millions of opportunities for entrapment. Because it is heavier than the other materials with which it is flowing, gold is pulled by gravity to the lowest possible level wherever it is found. Gold, therefore, lodges in hollow depressions, cracks and crevices in the bedrock of any waterway in which it is being carried. Tiny bits of the golden metal also become entangled in tree and grass roots and sink in the backwaters behind large boulders. The successful placer miner will soon learn to spot these traps and check them out with metal detector and gold pan.

Because fast-moving water has already done part of the work of sifting and sorting among rocks, gravel and gold, placer mining (panning) in streams is easier than dry panning. Of course, any type of panning is easier than hard rock mining which requires large amounts of equipment, plus the hard work of moving tons of rock to recover gold. This is why most recreational miners sell any hard rock claims they develop to large companies. It is much easier and more pleasant for the recreational miner to spend limited prospecting time in "working" with the more enjoyable opportunities of placer mining – especially with the advantages of electronic prospecting that are now available.

Dry Washes

Years of watching, first, movies and, now, television have given most people the idea that gold is found only in streams and rivers and only in deserted areas of the Western and Mountain States. *Not so!*

True, it is easier to recover gold if there is a good supply of running water, but water is never a necessity. The old-timers almost always followed the stream beds because they knew that if gold was there, it would be moved downstream by water from the mountains. And, early placer mining prospectors used this water to pan sand and gravel from heavier gold because it was the easiest way to find gold. Using today's modern gold pan, with its built-in riffle traps, a weekend prospector can find gold in a pan even without water.

What about the old dry washes? They were obviously once stream beds, but just as obvious is the fact that they have not contained water in centuries. Remember that these dry beds can contain gold, and that it became deposited in them just as it is now being deposited in streams of flowing water. Until the development of the modern metal detector, finding gold in dry areas presented the prospect of panning literally acres of dry ground. A discouraging proposition! With today's modern ground balanced metal detectors, prospecting is much easier, and the dry washes that were passed up during the last century can prove quite productive.

Productive also are old placer diggings. An early day placer miner could have been standing right on top of a large nugget and would never have spotted it unless it was gleaming brightly. Remember, he had only his eyesight to use as a locating tool.

Today's recreational miner can locate nuggets even though they are enclosed in rock or mud and buried under a layer of sand. By using the detector to scan the old diggings and placer mining areas, any gold that was simply bypassed can be found easily. In most instances, use of a correctly designed, quality metal detector is the only method by which these long-overlooked deposits can be found. In *all* instances, the use of a detector is the fast-

Facing
Roy Lagal searches with a small Super Sniper searchcoil to look between rocks beside a stream for gold or other prizes that were washed here by high water.

Over   Top
The Hand of Faith, 62-pound gold nugget, found in Australia by an amateur prospector using a Garrett detector, was sold for a reported one million dollars.

Bottom
Exposed tree roots such as these are often store-houses of nuggets and other pieces of ore that were captured here long ago.

Always look for gold **where** it has already been found. This rugged stretch of gold country is electronically prospected for ore and nuggets.

est, most practical way to pursue recreational mining. With a quality detector – designed for universal use like the Grand Master Hunter or, primarily for prospecting, the A3B-United States Gold Hunter – you can cover in minutes or hours what took months and years for the old prospectors who were limited to their shovels and pans.

Facing
After a detector locates a target in a stream, the prospector shovels material into a plastic gold pan to speed up nugget recovery.

Over
When Roy Lagal finds his target in these rocks, he will scoop it and surrounding material into the plastic gold pan he carries to locate it precisely.

Electronic prospecting can be combined with weekend travels and vacations. Just a short time spent by a stream can result in some nice color.

The beauty of this mountain stream is a joy that recreational prospectors experience often as they search for nuggets and placer deposits in gold country.

Hunting Float, Pockets

You have already learned that "float" can be defined as ore that broke off somehow from the mother lode and was carried away (usually down hill) by gravity, wind, water, earth movement or other acts of nature. Float can be found lodged in hollow places and behind boulders and other barriers to its downhill movement. These heavy pieces of ore form pockets on hillsides and have often been covered by an overburden of rock and gravel that has washed in on top of the gold ore.

Modern ground balanced detectors can locate these rich pockets of metallic ore. Medium to large searchcoils are preferable because they can search deeper and cover more ground. Both attributes will prove advantageous.

Begin your search for pockets and/or float on depressions in hillsides or in gullies, creek bottoms and other areas where you reason logically that heavy pieces of float might have stopped. Once you begin to locate float ore, work upstream or uphill from your initial discovery and try to locate the source. Remember that ore is a combination material and will not produce as strong a reading as that from a solid metallic object. Do not overlook any faint signals. When you get a reading over a pile of rocks, lay down your detector and proceed to move those rocks, *one at a time*, across the bottom of your searchcoil as you practiced in your bench tests. By following this procedure you can learn which rock(s) produced the signal.

Whether working in a creekbed or on a hillside, ground balance your detector carefully and sweep the searchcoil two to six inches above the surface, depending on the "chatter" you get from iron mineralization. Listen carefully for faint signals that can come from pockets deep under an overburden or those that have only a low percentage of metal. (Some iron mineral may also be present.) Nevertheless, if the pocket is relatively rich (conductive) and not too deep, you will mostly likely get a pronounced positive (metallic) response from your detector.

Chapter 12

A Rockhound's Metal Detector

I n the hands of a rockhound a quality ground balanced metal detector can be an invaluable aid, not only for helping identify specimens and determining conductivity, but for locating valuable samples. Although the detector can prove rewarding when properly used, nothing can replace the knowledge gained from experience in identifying semi-precious stones and gems.

The rockhound should consider the metal detector as equipment – even though, sometimes the most valuable – that can be used to locate and identify conductive metallic specimens that defy the naked eye. Specimens of the highest grade can often be overlooked, even by a trained eye. This is no wonder, since no eye can look *inside* a piece of rock the way that a detector can. Always remember that your detector is an instrument and a valuable one . . . but only when used properly and in the right location.

Become familiar with your metal detector. A whole new world will open up to you if you will just experiment. You will find you can spot-check promising rocks, and perhaps you will discover that those areas that are considered "worked out" may still produce promising specimens, after all.

Metal refers to any metallic substance of a conductive nature in sufficient quantity to disturb the electromagnetic field of a metal detector and produce a positive response. Gold, silver, copper and all the non-ferrous metals are just that – metals. *Mineral* refers primarily to minerals which produce a negative response from the metal detector. These are primarily magnetic iron and iron oxide (Fe_3O_4).

73

Conduct bench tests to familiarize yourself with the responses generated by various samples of which you already know the mineral content. Such testing will aid you greatly when you encounter specimens in the field whose mineral content you **do not know!** Test all likely-looking rocks. You'll find that only a few minutes work might produce a high grade specimen that has been passed over for years by fellow rockhounds.

The very low frequency of the modern detector's circuitry penetrates rocks easily and produces excellent results in identification of metallic ores. To identify *metal* vs. *mineral* specimens, use the Discriminate mode of your detector rather than its All Metal mode. Set your discrimination control at *zero* when you are certain that your detector has been calibrated at the factory to reject **no** metals at this setting. (See Chapter 8) You can now reject the specimens containing a high content of magnetite. Then, you may advance the discrimination setting slowly, just as you did in bench testing, to determine the content of the specimen and whether it is ferrous pyrite or a non-ferrous precious metal, such as gold, silver and copper.

Thorough and precise bench testing is a necessity if you are to be successful with a metal detector. Use of these procedures will quickly clear up confusion arising over specimens. Used in this manner your modern metal detector will provide you the full enjoyment of this wonderful hobby and keep you from missing valuable specimens.

When bench testing, always conduct each test in *exactly the same manner.* And, remember to test your specimen by moving it across your searchcoil as described for bench testing. Because of the winding configuration of co-planar or concentric searchcoils, your sample may respond differently on different areas of the searchcoil's surface.

For example, a large gold nugget placed on the *receiver* winding of a searchcoil will produce a positive (metallic) response. Place the same nugget on the portion of the searchcoil containing the transmitting coil winding, and you will receive a mineral (non-metallic) response. A small pebble of high grade magnetite iron ore will respond as mineral on a searchcoil's receiver winding and as metallic on the transmitting portion of that same searchcoil's winding. Concentric searchcoils, used by most manufacturers, do not have the above characteristics. Instead, they have a very uniform detection pattern over the entire bottom of the searchcoil. Always remember, pass the sample ACROSS the searchcoil at approximately 1 or 2 inches height.

Ask your metal detector dealer to demonstrate various marginal samples of easily recognizable gem or ore specimens. If the dealer admits a lack of knowledge concerning identification procedures, refer him to this book or some of the others published by Ram.

No matter how proficient you have proved yourself as a rockhound, do not dismiss lightly the capabilities of a modern metal detector. Just a small amount of some of the metals it will detect is worth a fortune. For example, consider the experience of one of the authors, Roy Lagal, at a major gem show. He encountered a lady rockhound selling specimens from a well known gold mine. She had retrieved discards from the mine's dump, broken them into baseball-sized pieces and priced them at $3 apiece. Upon just visual examination, Roy became curious and asked permission to test samples with his metal detector. All he tested responded as metal.

He explained his results to the lady and tried to explain the value of her "samples." She replied that she did not believe in metal detectors. To shorten this long story Roy purchased samples from her for $3 each, had them sawed into slabs, some of which were purchased by a jewelry maker for $125 each. Both of us have also high graded many silver samples that people had for sale. With a modern metal detector it is a simple matter to pick up the samples loaded with precious metal. This is just another way to make your detector pay off!

Conclusion

This book has truly been a labor of love for us . . . when we *first* wrote it almost a decade ago . . . each time that we revised and updated it . . . and, now, when we have completely rewritten the text to incorporate developments in metal detectors into the **21st Century.**

Our love, of course, is for electronic metal detecting generally and for prospecting in particular. Just the joy of being alive in the beautiful outdoors of gold country offers pleasure enough. To have the prospect of adding material wealth doing something we enjoy is akin to paradise.

Our labor consisted of long, hard hours spent in factory laboratories, at work benches and in the field testing, retesting and evaluating modern metal detectors of all types. We have proved many times just how valuable these detectors can be to the prospector and recreational miner. Without a doubt they spell the difference between success and a lack of it . . . the difference between a pleasant weekend getting exercise outdoors, or doing *exactly the same thing* at a personal profit of, perhaps, hundreds of dollars.

The one basic ingredient of success is the operator's ability to use a versatile modern ground balancing metal detector correctly. After you have studied this book thoroughly, we urge you to further your studies by reading several of the other volumes from Ram Publishing Company. Descriptions of these along with an order blank for them is contained in this book. They will round out your ability to understand and use the modern metal detector and the Gravity Trap gold pans.

The future of electronic prospecting is extremely promising. If you study this book and others, search out the gold fields and make a determined effort to apply in the field what you learned from our books, you cannot help but be successful. When you possess this knowledge, we are confident that you will be able to fill your poke many times with the earth's treasures. And, we also hope to . . . **see you in the field!**

Roy Lagal

Charles Garrett

Appendix

Basic sources for information on areas where gold might be found are given in this section. All of these addresses were current in 1988.

Forest Service Regions

Alaska Region
Federal Office Bldg.
P.O. Box 1628
Juneau, AK 99802

Intermountain Region
324 25th Street
Ogden, UT 84401

Northern Region
P.O. Box 7669
Missoula, MT 59807

Pacific Southwest Region
630 Sansome Street
San Francisco, CA 94111

Rocky Mountain Region
P.O. Box 25127
Lakewood, CO 80225

Southwestern Region
517 Gold Avenue SW
Albuquerque, NM 87102

Pacific Northwest Region
319 SW Pine Street
Portland, OR 97208

State Bureaus of Mines/Geology

Alaska Division of
 Geological Survey
3001 Porcupine Dr.
Anchorage, AK 99504

Arizona Geological Survey
845 N. Park Ave.
Tucson, AZ 85719

California Division of
 Mines and Geology
1516 Ninth St., 4th Fl.
Sacramento, CA 95814

Colorado Geological Survey
1845 Sherman, Rm 254
Denver, CO 80203

Idaho Geological Survey
Univ. of Idaho
Moscow, ID 83843

Montana Bureau of
 Mines & Geology
Montana College of
 Minl Sc & Tech
Butte, MT 59701

Nevada Bureau of Mines
 & Geology
University of Nevada
Reno, NV 89557-0088

New Mexico Bureau of Mines
 & Mineral Resources
Campus Station
Socorro, NM 87801

Oregon Dept. of Geology
 & Mineral Industries
1400 SW 5th Ave., Room 910
Portland, OR 97201-5528

South Dakota Geological
 Survey
Science Center, Univ. of S.D.
Vermillion, SD 57069

Utah Geological and
 Mineral Survey
606 Black Hawk Way
Salt Lake City, UT 84108

Washington Division of
 Geology & Earth
 Resources
Olympia, WA 98504

Geological Survey
 of Wyoming
P.O. Box 3008
University Station
Laramie, WY 82071

Bureau of Land Management
State Offices
(U.S. Department of the Interior)

Alaska
701 C St. Box 13
Anchorage, AK 99513

Arizona
P.O. Box 16563
Phoenix, AZ 85011

California
2800 Cottage Way
Room E-2841
Sacramento, CA 95825

Colorado
2850 Youngfield St.
Lakewood, CO 80215

Eastern States (All Others)
350 South Pickett St.
Alexandria, VA 22304

Idaho
3380 Americana Terrace
Boise, ID 83706

Montana (Also ND & SD)
P.O. Box 36800
Billings, MT 59107

Nevada
P.O. Box 12000
Reno, NV 89520

New Mexico (Also Kansas,
    Oklahoma and Texas)
P.O. Box 1449
Santa Fe, NM 87504

Oregon (Also Washington)
P.O. Box 2965
Portland, OR 97208

Utah
324 South State St.
Salt Lake City, UT 84111

Wyoming (Also Nebraska)
P.O. Box 1828
Cheyenne, WY 82003

BOOK ORDER FORM

Please send the following books

☐ **Modern Treasure Hunting**.................$12.95

☐ **Treasure Recovery from Sand and Sea**......$12.95

☐ **Modern Electronic Prospecting**...........$ 9.95

☐ **Buried Treasure of the United States**.......$ 9.95

☐ **Weekend Prospecting**.....................$ 3.95

☐ **Successful Coin Hunting**.................$ 8.95

☐ **Modern Metal Detectors**...................$ 9.95

☐ **Treasure Hunter's Manual #6**..............$ 9.95

☐ **Treasure Hunting Pays Off!**................$ 4.95

☐ **Gold Panning Is Easy**.....................$ 6.95

☐ **Treasure Hunter's Manual #7**..............$ 9.95

☐ **The Secret of John Murrell's Vault**.........$ 2.95

Garrett Guides to Treasure

☐ **Find An Ounce of Gold a Day**..............$ 1.00

☐ **Metal Detectors Can Help You Find Wealth**....$ 1.00

☐ **Find Wealth on the Beach**..................$ 1.00

☐ **Metal Detectors Can Help You Find Coins**...$ 1.00

☐ **Find Wealth in the Surf**....................$ 1.00

☐ **Find More Treasure With the Right Detector** $ 1.00

Ram Publishing Company
P. O. Drawer 38649
Dallas, TX 75238

Please add 50¢ for each book ordered (to a maximum of $2) for handling charges.

Total for Items	$_____
~~Texas Residents Add 8% Tax~~	$_____
Handling Charge	$_____
Total of Above	$_____
MY CHECK OR MONEY ORDER IS ENCLOSED.	$_____

I prefer to order through

☐ **VISA**   or ☐ MasterCard   (Check one.)

_____
Card Number

_____
Expiration Date

_____
Phone #

_____
Signature (Order must be signed.)

_____
NAME

_____
ADDRESS

_____
CITY, STATE, ZIP

CLEVENGER PAY-LES SALES
8206 North Oak
KANSAS CITY, MISSOURI 64118
1 800-999-9147